RETHINKING

ON

RELATIVITY

Hasmukh Rathod

Dedicated

To

My late father,

Who was the prime nurturer of my curiosity

Contents

Preface

Let me provide you the brief overview of this book. This book has contents as expected from the title of the book. The structure and foundation of the theory of relativity has been questioned by pointing out errors in the use of Lorentz transformation. The theory of relativity can be divided into two main parts, first is the special theory of relativity and second is the General theory of relativity. Both of these theories are criticized here. The special theory of relativity has two main postulates. Second postulate is about the observer specific constancy of the speed of light and the first postulates is about the application of laws of physics in inertial reference frames. It states that laws of physics are same in all inertial reference frames. In other word first postulates states that when two things are in a relative motion, then Lorentz transformation is applicable to find the relation of indices between two inertial reference frames. That Lorentz transformation is derived from the second and the first postulates of the theory by using simple geometry and calculations. The general theory of relativity also known as GR-General relativity is the expansion of the special theory of relativity and addition of postulates of curved four-dimensional spacetime. It deals with accelerated frames, Gravitational field, and co-ordinates applicable in curved geometry. GR states about behaviour of rods and clocks in the gravitational field and thus proposes a time dilation and length contraction phenomena in gravitational field.

Mistakes in the foundation of the theory of special relativity is suggested in relevant chapter. Derivation of the factor Υ using improper logic and math is the root of all misinterpretations. When we talk about the observer specific constancy of the light speed then it should lead to simple equations equally applicable for two different situations. 1. When two object moves toward each other and when two objects move away from each other. That should show both compensatory time dilation and time contractions. But Tricky use of factor gamma and removing second observer by violation boundaries of the Lorentz transformation for a time dilation calculations resulted in time dilation in both these situations. this violation of mathematics is described in relevant chapter. It is not possible to reject a preferred reference frame using special relativity. Limitation of the maximum speed of matter and light points out the existence of something that restricts the free motions without limits in the space. This topic is covered well in this book.

Mistakes in the calculation of Lorentz Fitzgerald length contraction is also highlighted and correction is suggested. Photon clock examples provide the better understanding of time dilation and time contraction based on the observer specific constancy of the light speed.

So, in the for-coming chapters, maths and sound logic are used to rethink on the relativity theory.

Dr. Hasmukh Devidas Rathod

(M.B.B.S)

Acknowledgments

This book is a result of the learning from the various media including websites in a public domain on internet and You-tube. I thanks to the authors who shared their knowledge and information on public domains. I thanks to the Wikipedia for providing a wast platform for studying various topics.

1. Twin paradox

Interchangeability of subjects in a special theory of relativity.

In special theory of relativity by albert Einstein, there was no universal reference frame accepted. There is a concept of the preferred moving inertial reference frame which was introduced later to deal with the twin paradox.

According to the first postulate of the special theory of Relativity all inertial reference frame should have same and simple law of physics within a frame. when two observers A and B move in relation to each other, For A, itself is steady and B is moving, similarly for B, itself is steady and A is moving. If we apply Lorentz transformation, there is slowing down of the time of B for A and similarly there should be slowing down of the time of A for B.

This subject interchangeability causes paradoxes, and it suggests nonrealistic impossible outcomes.

I would like to explain it with example of uncle -nephew relationship.

U is uncle of N

N is nephew of U.

What subject interchangeability states is that application of this relation is same for each of the two subjects in relation to each other.

Means it states that,

For U, U is an uncle of N and N is nephew of U.

And For N, N is an uncle of U and U is a nephew of N.

Means both feels own self as uncle of the other one.

Now is it clear that such relation is impossible, which can be seen in the twin paradox.

Let me explain you twin paradox with the help of an example of two clocks.

S and M are two clocks.

Both are moving in relation to reach other.

For S, S itself is steady and M is moving so, time in M will be slow down.

For M, M itself is steady and S is moving so time in S will be slow down.

So according to the Special theory of Relativity with subject interchangeability, M shows less time than S and S also shows less time than M. We can understand that this is not possible. This is a twin paradox.

Suppose S and M are two twin brothers.

When S is on the earth and M is going on a space trip with tremendous speed, and came back to the earth, there would be a relative time dilation for each subject.

S will find that he is aged to an adult and M is still a child because for S the relatively moving subject was M which should show a slowing down of time.

Similarly, M will find that he is aged to an adult and S is still a child because for M the relatively moving subject was S which shows a slowing down of time.

For S there should be slow ageing of M and for M there should be slow ageing of S.

When they both meets, there is an impossible controversial situation were finding of both subjects would be personal and impossible. It is not possible that two persons can be elder than each other simultaneously.

This was considered as the error in the Special theory of Relativity by few peoples, but Einstein tried to explain the situation and favors a preferred moving inertial reference frame based on situation and presence of the acceleration experienced by subjects prior or during the journey. He stated that which one feels acceleration is moving and which one don't feel acceleration is steady and time dilates for moving in relation to steady. So, here that who goes to the space trip will show a time dilation in relation to steady subject. So, in above example S will age to an adult but M won't be due to time dilation for moving subject only.

We know that acceleration is felt for a time when object achieves speed, after achieving speed if it moves with constant velocity then it feels no more acceleration. What factor affects during that constant velocity motion which causes time dilation even after acceleration is no longer remained? This suggests existence of something that decides time dilation in relation to it. That something is not mentioned in the Special theory of Relativity.

Also, there is another way to decide Preferred reference frame which is a type of relative motions. Suppose if an asteroid orbits a planet, then both planet and asteroid are moving in relation to each other but by considering an apparent relative motions of the distant stars and analyzing a motion in relation to each other it would become evident that asteroid is orbiting a planet and not a planet is showing an inconsistent variable motion in relation to asteroid. So, with all this application special theory of the relativity states that there could be a time dilation in a moving reference frame and not in a nonmoving one. When both are feeling acceleration then there would be a complex relative relation which would decide time dilation more or less in relation to each other depending on their movements considered with effect of acceleration.

All this suggests that the Special theory of Relativity is not viable without determining motions with effect of accelerations. Application of time dilation with subject interchangeability is not possible and one has to define motions by considering relation with the rest of the world.

2. Preferred frame of reference in the special theory of relativity

It is believed that there is a no preferred frame of reference for the relative motions of two objects in special relativity theory. Really is it so? Are two objects free to move with any speed in relation to each other? If three objects are there, are they free to move with any speed in relation to one another within a range of speed from 0 to the speed of light c? Answer is no. They can't move with speed beyond some limit in relation to one another according to the special theory of relativity. Though there is a no preferred reference frame but still there is a range that restricts motions of all objects. Let's see a thought experiment for these.

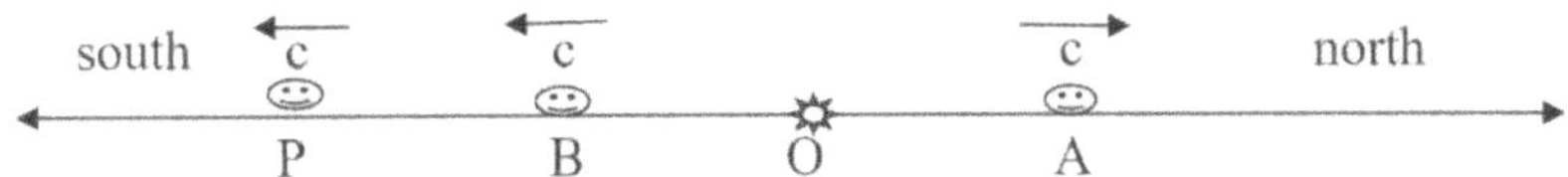

Figure: 2.1

Suppose there are four objects as well as observers named A, O, B and P. We would like to consider O as a steady observer (not feeling acceleration during this experiment). Now speed of the A in relation to O can't be more that c, the speed of light. Suppose A is moving with the speed c in a direction of north relative to O. B is also moving with the speed c in opposite side in a direction of south in relation to O. For the observer A, O would be stand still in the time and it won't perceive any change in O, what could be the observation of B from the observer A? That is an impossible physical system according to special theory of relativity. Let's clear it further. Here B is moving with the speed c in relation to O. Now, consider P as an observer moving with the speed of c in relation to B in a same direction of the motion toward which B is moving. As per the classical physics speed of P relative to O would be 2c, but in the context of the special theory of the relativity relative, speed more than c can't be possible. So, adding P with a speed c relative to B would force a situation to change in a situation where relative speed of B in relation to O must not be c or it should be less than c.

This concludes that we can't add series of objects in a line with speed c relative to the nearby object. This would violate one of the postulates of the special theory of the relativity.

Note: Anything which moves with a relative speed equal to c must lose its mass nature and become energy or radiation wave form. This phenomenon is not taken in a consideration during this thought experiment shown above to understand the point of the thought experiment.

We can say that theory of special relativity concludes that there could not be the existence of mass or anything which could travel with the speed more than that of the light in relation to observer and there could not be addition of any object 2 moving away from the observer also moving with any speed more than zero in relation to the object 1 which is moving with the speed of light relative to the observer. Here all motions are to be considered in a single linear direction. Thus, a physical system contains something with restricted motions in relation to one another. It seems like that there is the environment adhered to the physical system that restricts the relative motions beyond some limit. That environment doesn't allow limitless speed of motions.

3. Observer specific constancy of the speed of light

You There was a complex observation in a Michaelson Morley experiment which was intended to find a drag by aether on the light propagation.

Observer specific constancy of the speed of light is not easy to understand. Actually, it is so because it is a poorly defined concept itself.

Look at the figure 3.1below.

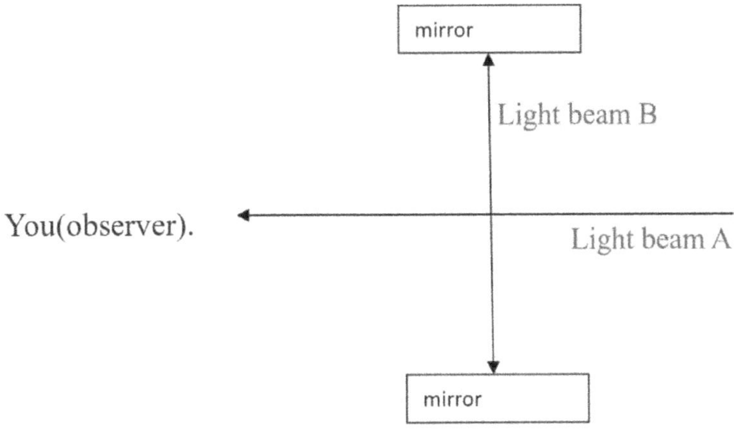

Figure: 3.1

There are two beam of the lights A and B. The whole assembly is put in a ship moving with a velocity v in relation to you in horizontal direction.

According to the special theory of relativity there is a relative time dilation for that ship due to its relative motion in relation to you. There would be some different time environment in the ship relative to you. Now both beams of the light are also moving with the ship in relation to ship and also in relation to you. Let me discus motion in relation to you only. A is travelling in a direction toward you with the speed of light. Its speed relative to you must be c. Now B is also traveling with the speed c in relation to you according to the Special theory of Relativity. Because path of B would be of zig-zag pattern due to bouncing of light beam between those two mirrors shown in the figure. After a passage of time when ship travels some distance then light beam B will remain at same place bouncing between two mirrors near to beam A. Here if the speed of light of the B would be equal to A for you than it must be left behind during this travel. But that doesn't happen. Relativity may explain this by saying that there is a time dilation when you see bouncing beam B and there is a no time dilation when you see beam A. Due to direction of motion this thing occurs. But we can see here clearly that light beam B is travelling faster in relation to light beam A to stay in the assembly between those two mirrors along with light beam A. This can be compared to the arrangement with interferometer in a Michaelson Morley's experiment where in relation to the space two beam of light travels in two different directions where beam traveling in a zig-zag path travels more distance in relation to space compared to beam traveling in a straight line. After some time both light beams in the interferometer meet at a point which is at equal distance for paths of lights in relation to interferometer from the source of light in a Michelson Morley's experiment and forms the interference. So, where is the observer specific constancy of the speed of light here? It is evident that speed of the assembly is being added to the vertical beam and it get dragged with the assembly. Obviously, there is a difference in the speed of light for both beams of light. This raises questions against observer specific constancy of the speed of light in a vacuum.

So, conclusion is that there can be an addition to the speed of light if it is hit enough to deviate it from its path. Its speed would remain constant when it is not getting momentum that deviates it from its straight path, when it gets an additional speed in the direction of the deviation or drag. The total speed can be calculated from the vector addition operation. Addition of the speed could be in a perpendicular direction

4. Mathematical mistakes in Special Relativity

There are some internal conflicts and errors while deriving time dilation equation and getting co-ordinate transformation with Lorentz's transformation as described below.

The Special theory of Relativity has the two basic postulates which has potential to derive other predictions by using math with those postulates.

Let's see those two basic postulates of special theory of relativity.

1st: - Laws of physics are same and can be stated in their simplest form in all inertial frames of reference.

2nd: - Speed of light is constant for the observer irrespective of the reference frames and relative speed of the source of light.

Lorentz transformation is a method for the co-ordinate transformation in the special theory of relativity that replaces the Galilean transformations of classical physics.

I want to point out the mistake in this Lorentz transformation method.

Let's see calculations of Lorentz transformation equation with simple method.

On the next page, there is a figure 4.1 shown to explain Lorentz transformation.

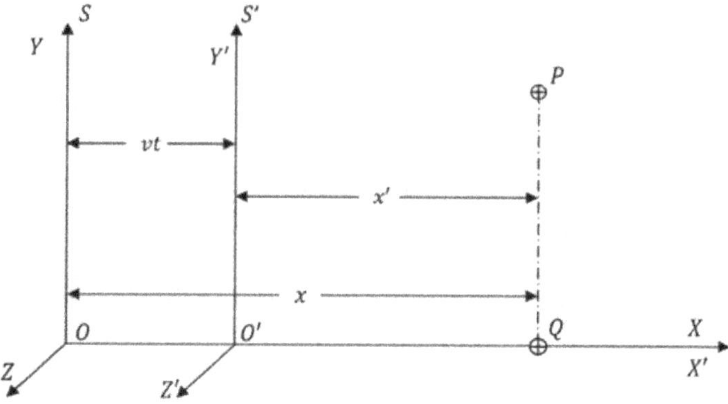

Figure: 4.1

Here two inertial reference frames S and S' are aligned with cartesian planes, which defines co-ordinates of the point $P(x, y, z, t)$ and $P'(x', y', z', t')$ respectively in S and S' inertial reference frames. Here S' inertial reference frame is moving with uniform rectilinear velocity "v" along the direction of positive X axis of S frame. There is no movement of S' along Y or Z axis of the frame S. Both frames are aligned in a way that O and O', co-insides at time $t = t' = 0$. Here P is the point selected randomly in the stationary reference frame S toward Positive X axis side of Cartesian plane. Q is the reference point on X axis in the reference frame S with co-ordinates $(x, 0, 0, t)$. In a time "t" photon travels perpendicular to Y axis from the point P for a distance equal to $OQ = x = ct$ and meanwhile during the same time t, S' frame moves a distance vt along positive X axis.

Note:

(One thing to-be noted here is that, though we are transforming co-ordinates for P, we are doing all calculations for Q. Understand this, for the conveniency of the calculations we have assumed that $y' = y$, and $z' = z$. Means relative movement of P for S' frame is only on X axis. So, there will not be any change in y and z during transformations, so they will not appear in calculations, only x will appear. We will define a time co-ordinate for P in a way that x will be equal to ct. Here OP is not equal to ct, but OQ is equal to ct and all calculations are for Q and it will be applicable as a calculation of x co-ordinate of P while transformations.

We will not use confusing relation as below,

$$x^2 + y^2 + z^2 = d^2 \qquad here\ d = OP = ct$$

$$x^2 + y^2 + z^2 =$$
$$c^2 t^2 \quad \dots \dots \dots \dots \dots \dots \dots \dots \dots \dots \dots (C)$$

Why are we not taking these relations?

Because here we are able to calculate for x only and we will have to take $y = 0, z = 0$ to derive constant Y.

When we do that, these relations would become,

$$x^2 + (0)y^2 + (0)z^2 = c^2 t^2 \ ,$$

$$x^2 = c^2 t^2$$

So,

$$x$$
$$= ct \qquad (no\ negative\ value\ for\ time.)$$

If we put this value in equation C,

$$c^2 t^2 + y^2 + z^2 = c^2 t^2$$

It will yield $y = 0$, and $z = 0$, means we are restricted to choose Point on the X axis only for calculations, after getting transformation for x, we could apply it to co-ordinates of P in S' frame, but the equations for calculation of x, must consider $y = 0$, and $z = 0$, otherwise that won't work.

So, Lorentz transformation equations are for Q and not for P. We just apply transformation of co-ordinates of Q to the x co-ordinate value of P. So, there will be Q only in calculations ahead.

Relation between O and Q is same as relation between P and nearest point on Y axis of S frame.

Relation between O' and Q is same as relation between P and nearest point on Y' axis of S' frame.

So, $x = ct$ is applied wherever there was a need in calculations,

We are free to choose a point on the X axis so, there should not be an objection for calculations for the point Q.)
End of the note.

From the figure 4.1

x is the co-ordinate of P on the X axis.

$$x = ct \qquad \text{...... (1)} \qquad \text{(read "note" above)}$$

And

$$x' = ct' \qquad \text{...... (2)} \qquad \text{(as per the postulate 1 of SR)}$$

From the figure 4.1, Newtonian relations can be,

$$x' = x - vt \qquad \text{...... (3)}$$

and

$$x = x' + vt \qquad \text{...... (4)}$$

If we apply relativistic relation in eq.3,

$$x' = Y(x - vt) \ldots \ldots \ldots \ldots \ldots \ldots \text{.. (5)}$$

Here Y is the constant.

Now using first postulates of the special theory of relativity.

(using subject interchangeability)

$$x = Y(x' + vt') \dots \dots \dots \dots \dots \dots (6)$$

I have an objection for this equation 6 and this is the root for all errors in the Special theory of Relativity. Applying such constant is not logical.

For now, continuing the calculation.

Putting the value of x' in the equation no (6) from the equation no (5)

$$x = Y(x' + vt')$$

$$x = Y(Y(x - vt) + vt')$$

$$x = Y^2 (x - vt) + Yvt'$$

$$x = Y^2 x - Y^2 vt + Yvt'$$

$$Yvt' = x - Y^2 x + Y^2 vt$$

$$t' = \frac{x - Y^2 x + Y^2 vt}{Yv}$$

$$t' = \frac{x - Y^2(x - vt)}{Yv} \dots \dots \dots \dots \dots \dots (7)$$

Now using second postulates of special theory of relativity.

Distance OQ is Equal to x and is equal to ct

Distance $O'Q$ is equal to x' and is equal to ct'

So,

$$x = ct$$

$$x' = ct' \quad \text{........ ....} \ (2)$$

Now putting the value of x' & t' in this equation (2)

$$x' = ct'$$

$$Y(x - vt) = c.\frac{x - Y^2x + Y^2vt}{Yv}$$

$$Yx - Yvt = c.\frac{x - Y^2x + Y^2vt}{Yv}$$

$$Yx * Yv - Yvt * Yv = c(x - Y^2x + Y^2vt)$$

$$Y^2xv - Y^2v^2t = cx - Y^2cx + Y^2cvt$$

Now putting the value of $x = ct$ in the above equation

$$Y^2xv - Y^2v^2t = cx - Y^2cx + Y^2cvt$$

$$Y^2ctv - Y^2v^2t = c * ct - Y^2c *$$
$$ct + Y^2cvt$$

$$Y^2ctv - Y^2v^2t = c^2t - Y^2c^2t + Y^2cvt$$

Cancel Y^2cvt on both sides.

$$-Y^2v^2t = c^2t - Y^2c^2t$$

$$c^2t = -Y^2v^2t + Y^2c^2t$$

$$c^2t = Y^2c^2t - Y^2v^2t$$

$$c^2t = Y^2t(c^2 - v^2)$$

$$c^2 = Y^2(c^2 - v^2)$$

$$Y^2 = \frac{c^2}{c^2 - v^2}$$

Dividing above and below with c^2

$$Y^2 = \frac{1}{1 - v^2/c^2}$$

$$Y = \frac{1}{\sqrt{1 - v^2/c^2}}$$

Now putting the value of Y in the equation no (7),

$$t' = \frac{x - Y^2(x - vt)}{Yv} \quad \ldots \ldots \ldots \ldots \ldots \ldots \ldots \; (7)$$

$$= \frac{x - \dfrac{c^2}{c^2 - v^2}(x - vt)}{v\dfrac{c}{\sqrt{c^2 - v^2}}}$$

$$= \frac{\dfrac{c^2 x - xv^2 - c^2 x + c^2 vt}{c^2 - v^2}}{\dfrac{vc}{\sqrt{c^2 - v^2}}}$$

$$= \frac{c^2 x - xv^2 - c^2 x + c^2 vt}{vc\sqrt{c^2 - v^2}}$$

$$= \frac{-xv^2 + c^2 vt}{vc\sqrt{c^2 - v^2}}$$

$$= \frac{c^2 vt - xv^2}{vc\sqrt{c^2 - v^2}}$$

$$= \frac{c^2 v(t - xv/c^2)}{vc\sqrt{c^2 - v^2}}$$

$$= \frac{c(t - xv/c^2)}{\sqrt{c^2 - v^2}}$$

$$= \frac{t - {xv}/{c^2}}{\frac{\sqrt{c^2 - v^2}}{c}}$$

$$= \frac{t - {xv}/{c^2}}{\sqrt{\frac{c^2 - v^2}{c^2}}}$$

$$t' = \frac{t - {xv}/{c^2}}{\sqrt{1 - \frac{v^2}{c^2}}} \qquad \dots\dots\dots\dots\dots (8).$$

Here $v < c$.

So, Lorentz transformation equations are,

$$x' = \frac{x - vt}{\sqrt{1 - \frac{v^2}{c^2}}} \dots\dots\dots\dots\dots\dots\dots (9)$$

$$y' = y$$

$$z' = z$$

$$t' = \frac{t - {xv}/{c^2}}{\sqrt{1 - \frac{v^2}{c^2}}}$$

Inverse Lorentz transformation equations would be,

$$x = \frac{x' + vt'}{\sqrt{1 - \frac{v^2}{c^2}}} \qquad \dots\dots\dots\dots\dots\dots\dots\dots (10)$$

$$y = y'$$

$$z = z'$$

$$t = \frac{t' + \frac{xv}{c^2}}{\sqrt{1 - \frac{v^2}{c^2}}}$$

What is the postulate 2 used here? it is the observer specific constancy of the light speed. Here we have considered only one situation where there is a motion of the moving reference frame toward the source of the light. What if that reference frame would move away from the source of the light. There would be a negative velocity, and these would result in following equations for the figure 4.2 shown below which would be as inverse Lorentz transformations equations.

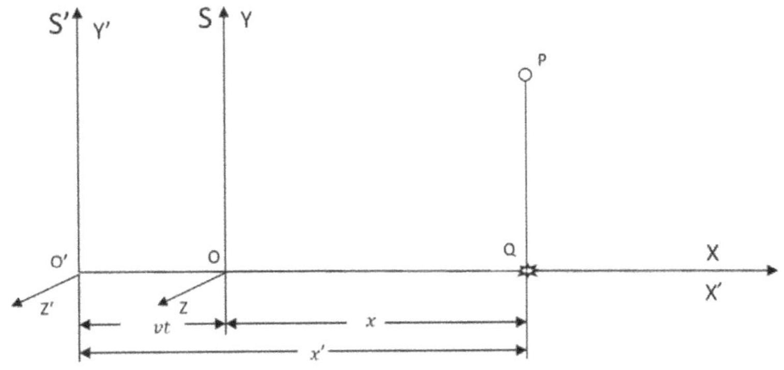

Figure: 4.2

Here only difference is S' is moving away from the P and Q.

Relations would be like,

$$x' = Y(x + vt) \ldots\ldots\ldots\ldots\ldots\ldots(11)$$

Here Y is the constant.

Now using first postulates of the special theory of relativity.

(using subject interchangeability)

$$x = Y(x' - vt') \dots \dots \dots \dots \dots \dots (12)$$

This would result in,

Transformation equations,

$$x' = \frac{x + vt}{\sqrt{1 - \dfrac{v^2}{c^2}}}$$

$$y' = y$$

$$z' = z$$

$$t' = \frac{t + {xv}/{c^2}}{\sqrt{1 - \dfrac{v^2}{c^2}}}$$

These are same as inverse Lorentz transverse equations.

Remember my objection for the equation no 6

That objection intended to say that it is a root for errors and paradoxes. If we are trying to find relativistic relation by applying observer specific constancy of the speed of light, then adding a constant Y is wrong logically. There should be following relations only.

$$x' = x - vt \dots \dots \dots \dots \dots \dots (3)$$

and

$$x = x' + vt' \dots \dots \dots \dots \dots \dots (4)$$

Now,

$$x = ct$$

And

$$x' = ct' = x - vt$$

$$ct' = ct - vt$$

$$t' = \frac{ct - vt}{c}$$

$$t' = t\frac{c - v}{c}$$

Now

$$x' = ct'$$

put the value of x' & t' in this equation

$$x' = ct'$$

$$x' = c.t\frac{c-v}{c}$$

$$x' = t(c-v)$$

Here $v < c$.

So, relativistic transformation equations are,

$$x' = t(c-v) \quad \dots\dots\dots\dots\dots\dots (13)$$

$$y' = y$$

$$z' = z$$

$$t' = t\frac{c-v}{c} \quad \dots\dots\dots\dots\dots\dots (14)$$

If moving reference frame is moving away from the source of the light then,

relativistic transformation equations would be,

$$x' = t(c+v)$$

$$y' = y$$

$$z' = z$$

$$t' = t\frac{c+v}{c}$$

This would show a time contraction.

So, in a case where something moves away first, then came back, as a sum of event there would be following change in a relative time.

$$2t' = t\frac{c+v}{c} + t\frac{c-v}{c}$$

$$2t' = t\frac{c+v+c-v}{c}$$

$$2t' = t\frac{2c}{c}$$

$$2t' = 2t$$

$$t' = t$$

So, when object comes back there would not be any net time dilation or time contraction if we follow postulate 2 based relativity here as shown above.

So, really observer specific constancy of the light doesn't yield any absolute time dilation, that is a wrong calculation of the constant Y that results in such wrong calculation of the time dilation. Mathematics doesn't support it.

Let's see why.

Following equation gives the value of dilated time in moving reference frame,

$$t' = \frac{t - \frac{xv}{c^2}}{\sqrt{1 - \frac{v^2}{c^2}}} \qquad \ldots\ldots\ldots (8)$$

Time dilation:
If we put a clock at the origine of the cartesian plane attached to the moving S' reference frame, then x would be Zero.
Now put $x = 0$ in the equation 8.

$$t' = \frac{t - \frac{(0)v}{c^2}}{\sqrt{1 - \frac{v^2}{c^2}}}$$

$$t' = \frac{t - 0}{\sqrt{1 - \frac{v^2}{c^2}}}$$

$$t' = \frac{t}{\sqrt{1 - \frac{v^2}{c^2}}}$$

$$t' = Yt$$

So, if the clock moves with a speed equal to v in relation to observer, time t' would be slowed down in a clock as per the equation $t' = Yt$, which is considered as a relativistic time dilation.

This was calculations according to special relativity.

Criticism:

Criticism About time dilation.

In figure 4.1 if we want to put a clock at the origin of the S' frame.

Then value of the x becomes zero and we get time dilation equation as above.

But we have used in the derivation of the Lorentz transformation that $x = ct$.

So, if we take $x = 0$ then ct becomes zero.

If $ct = 0$ then t must be 0 because c is the constant and it can't be zero.

If t is zero, then t' also becomes zero and no event can happen in a figure 4.1 and all objects remain aligned on one vertical line on Y axis. Time dilation can't be defined in this situation.

So, showing time dilation by taking $x = 0$ is violation of the laws of the mathematics. Even without mathematics if we try to understand that x is a distance that light travels in a time t' and if a distance is zero then time taken for the light to travel must be zero. If we don't consider here t' equal to zero than we are trying to say that light takes more time to travel a no distance! So, assuming a clock on the point of observer is violating requirements for the Lorentz transformation and it don't yield anything logically sound.

Similarly, calculation of Lorentz Fitzgerald length contraction is also wrong. Within the context of the special theory of relativity also and there should be new calculation if we reject the derivation of the constant Y.

What this means is that time dilation and length contraction are relative thing even within the scope of special relativity and Lorentz transformation and those can be appreciated only when two observers observe object away from them. Superimposing that object on any one of the two observers violates mathematics. If we reject the derivation of the constant Y and accept simple relativistic equations for observer specific constancy of the light, there could be a derivation of relative time contraction and relative length dilation also.

Photon clock showing just a relativistic time dilation is also not justifiable because if we change the arrangement in a photon clock it becomes able to show relative time contraction also. Photon clock type one (used to explain relativistic time dilation) and photon clock type two are explained ahead in this book.

So, Major mistake is accepting following relations true.

$$x' = Y(x - vt) \dots \dots \dots \dots \dots \dots (5)$$

$$x = Y(x' + vt') \dots \dots \dots \dots \dots \dots (6)$$

This is the mistake at a root. Derivation of factor Y is wrong.

Another mistake is violating limitations of the Lorentz transformation to derive Time dilation and length contraction.

5. Showing limitations of the Lorentz transformations with its attributes

Widely misused calculator is a Lorentz transformation. It is a calculator that calculates or transforms co-ordinates of the point between two relatively moving inertial reference frames with uniform rectilinear velocity in relation to each other. This calculator shows its works at the end of the event. Means when light reach from event point to the origin of each moving reference frame. So, in a time t, when light from the event point reach to the origin of each reference frame, a moving reference frame moves a distance vt. This is the basic understanding of the situation used in the Lorentz transformation. This confirms the restrictions shown in this topic here. The Lorentz transformation is applicable only with these restrictions.

Lorentz transformation is a widely misused tool to derive mysterious outcomes by crossing its limitations. Here with I am going to set those limitations because it is evident from the concept of the mechanism of the tool.

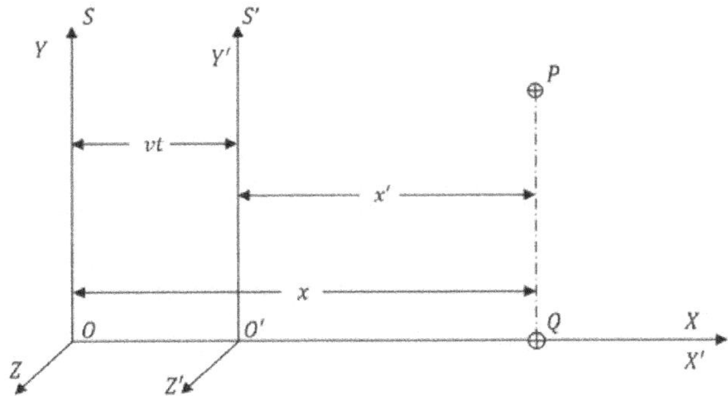

Figure 5.1

Cartesian planes are attached to S and S' inertial reference frames. S' inertial reference frame is moving with uniform rectilinear velocity "v" along with positive X axis.

Point P is chosen randomly on the side of positive X axis in a stationary reference frame S,Co-ordinates of the P in the S frame is (x, y, z, t) and co-ordinates of the point P in the S' frame is (x', y', z', t'). Q is a point on the positive X axis with co-ordinates $(x, 0, 0, t)$

Here, P is on the S, stationary and non-primed frame.

Here it has been said that S' moves along with X axis only, means y and z co-ordinates are not changed during relative motion. From the figure 5.1 it is evident here that, change is there in x and t co-ordinates only. These two co-ordinate changes in same manner for P and Q. If we just calculate these two-co-ordinate transformation for Q then it will be applicable to the P also. So, we will do that only because that is enough.

Here there is an Event at Q. light arises from Q and reaches to origin of both frames S and S'.

S' is moving, it is moving with the constant uniform rectilinear velocity v along the X axis of the S frame.

So, in a time t, light will travel a distance $x = ct$ and will reach to origin of S, mean while in a reference frame S' light will travel a distance x' in a time t' and will reach to origin of the reference frame S'. During this period, moving reference frame S' will travel a distance vt along with the positive X axis of the S frame resulting in the following relations.

$x = ct$ and $x' = ct'$

Visible relations are,

$$x = x' + vt'$$

And

$$x' = x - vt$$

Relativistic relations will be,

$$x = Y(x' + vt')$$

And

$$x' = Y(x - vt)$$

this will result in a Lorentz transformation equations as follows.

$$x' = \frac{x - vt}{\sqrt{1 - \frac{v^2}{c^2}}}$$

$$y' = y$$

$$z' = z$$

$$t' = \frac{t - \frac{xv}{c^2}}{\sqrt{1 - \frac{v^2}{c^2}}}$$

And Inverse Lorentz transformation equations as follows.

$$x = \frac{x' + vt'}{\sqrt{1 - \frac{v^2}{c^2}}}$$

$$y = y'$$

$$z = z'$$

$$t = \frac{t' + \frac{x'v}{c^2}}{\sqrt{1 - \frac{v^2}{c^2}}}$$

So, in Lorentz transformations,

Followings are limitations,

$x \neq 0$, value of the x must be more than zero, because when we assume event on the Y axis, it will take a no time for a light to reach Y axis from the event point. S' moves with a velocity v for the time t which is time taken by light for traveling the distance x. If the distance x is zero, then time t must become zero because x is the distance travelled by light with the speed c for a time t and c is the constant which can't be zero. That's why t must be zero here. In that "no time" S' won't be able to move, and there will not be a generation of relations required to derive equations of transformations, and Lorentz transformation won't work at all.

Now, for that nonzero x, distance ct can't be 0, so, it leads to $t \neq 0$, that will further lead to $vt \neq 0$.

Distance travelled by S' frame must be more than 0, because for any $x \neq 0 \Leftrightarrow t \neq 0 \Leftrightarrow ct \neq 0 \Leftrightarrow vt \neq 0$. So, in a situation described above, for any $x \neq 0 \Leftrightarrow t \neq 0$, value of vt can't be 0.

There must be,

$v < c$, (if we take $v = c$ or $v > c$, then value of the Y will become undefined)

$t > 0$. (If we take $t = 0$, then $x = ct = 0$, leads to no event and no calculations.)

In Lorentz transformation there are inter dependency of indices. x is the distance travelled by light with the speed c and in a time t. Here the speed of light c is the constant and does not change. For a specific value of the distance x there would be specific value of the time t. Reference frame S' travels a distance vt in a time t with the velocity v which is not zero. For the specific x there would be specific t and there would be specific vt also. when we choose two different values of x, there would be different specific values of t, and vt would result according to value of x. That means,

If $x_1 \neq x_1$,

Then, $t_1 \neq t_1$, $vt_1 \neq vt_1$.

Means,

$$x_1 \neq x_1 \Leftrightarrow t_1 \neq t_1 \Leftrightarrow vt_1 \neq vt_1$$

So, the Lorentz transformation is applicable only with following limitations,

1. $v < c$

2. $t > 0$

3. $x \neq 0 \Leftrightarrow t \neq 0 \Leftrightarrow ct \neq 0 \Leftrightarrow vt \neq 0$.

4. $x_1 \neq x_2 \Leftrightarrow t_1 \neq t_2 \Leftrightarrow ct_1 \neq ct_2 \Leftrightarrow vt_1 \neq vt_2$.

Violation of these **attributes** is violating a mathematics.

Lorentz transformation equations are valid only when the limitations of the Lorentz transformation taken in considerations. Situations that cross these limitations don't fit in Lorentz transformation, there must be a fresh transformation calculation for such situations and direct use of Lorentz

6. mistakes in the calculation of the Lorentz Fidge Gerald's length contraction

Let me show you mistakes in the calculation of the Lorentz Fidge Gerald's length contraction.

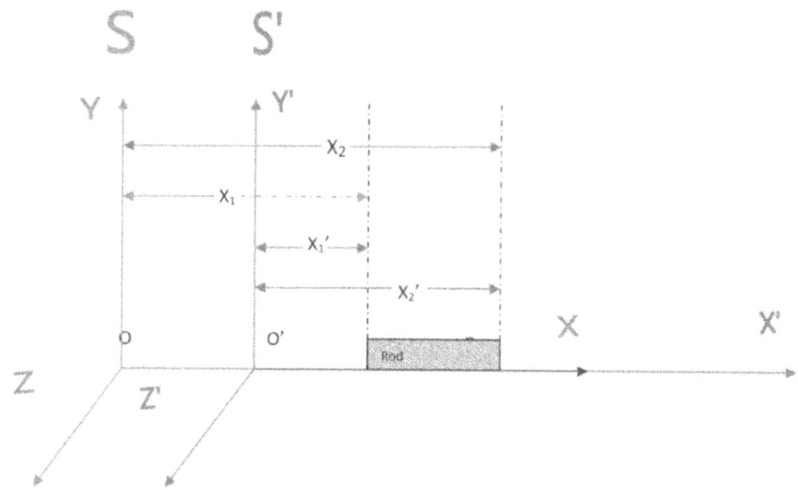

Figure: 6.1

S' inertial reference frame is moving with uniform rectilinear velocity "v" along the X axis.

Rod is placed on the positive side of the cartesian plane aligned to the X axis of the S inertial reference frame. In stationary inertial reference frame S, distance of proximal end of Rod from the origin of the reference frame is x_1 and distance of the distal end of the rod is x_2, Proper length of Rod in S is l_0.

In moving inertial reference frame S', distance of proximal end of Rod from the origin of the reference frame S' is x_1' and distance of the distal end of the rod is x_2'. So, contracted length of the Rod in S' is l.

Derivation of the Lorentz Fitz Gerald length contraction using Lorentz transformation equations is done in the special theory of the relativity as follows.

Lorentz transformation equation for x co-ordinate is,

$$x' = \frac{x - vt}{\sqrt{1 - \frac{v^2}{c^2}}} \quad \dots \dots \dots \dots \dots \dots \dots \dots \dots \dots \dots (9)$$

$$x = \frac{x' + vt'}{\sqrt{1 - \frac{v^2}{c^2}}} \quad \dots \dots \dots \dots \dots \dots \dots \dots \dots \dots (10)$$

In a figure 6.1, a rod is put along the X axis. Length of the rod in relation to stationary reference frame is proper length l_0 Length of rod in relation to moving reference frame is contracted length l

Proper length $\qquad l_0 = x_2 - x_1 \qquad \dots \dots \dots (15)$

Contracted length $\qquad l = x_2' - x_1' \qquad \dots \dots \dots (16)$

Now let put value of (10) in the equation no (15)

$$l_0 = x_2 - x_1$$

$$l_0 = \frac{x_2' + vt_2'}{\sqrt{1 - \frac{v^2}{c^2}}} - \frac{x_1' + vt_1'}{\sqrt{1 - \frac{v^2}{c^2}}}$$

$$= \frac{x_2' + vt_2' - x_1' - vt_1'}{\sqrt{1 - \frac{v^2}{c^2}}}$$

But, both length x_2' and x_1' are measured simultaneously at a time t' in moving inertial reference frame S'.

(note: this line is used to derive length contraction traditionally, though this is not possible, but we continue here to see calculations)

So, $t_1' = t_2' = t'$

So,

$$l_0 = \frac{x_2' + vt' - x_1' - vt'}{\sqrt{1 - \frac{v^2}{c^2}}}$$

$$l_0 = \frac{x_2' - x_1'}{\sqrt{1 - \frac{v^2}{c^2}}}$$

$$l_0 = \frac{l}{\sqrt{1 - \frac{v^2}{c^2}}} \qquad \ldots\ldots\ldots\ldots\ldots\ldots\ldots (17)$$

And,

$$l = l_0 \sqrt{1 - \frac{v^2}{c^2}} \qquad \ldots\ldots\ldots\ldots\ldots\ldots\ldots (18)$$

Here we can see that for described value of v, l will always be less than l_0

Because value $\sqrt{1 - \frac{v^2}{c^2}}$ will always be less than 1 for $0 < v < c$ value of v.

So, Length contraction is Δl

$$\Delta l = l_0 - l$$

$$\Delta l = l_0 - l_0 \sqrt{1 - \frac{v^2}{c^2}}$$

$$\Delta l = l_0 \left(1 - \sqrt{1 - \frac{v^2}{c^2}}\right)$$

Note: Here I have shown a mistake in the logic applied by putting a note in a bracket, but here the calculation of Lorentz Fitzgerald length contraction is shown as it is described in the special theory of relativity by Professor Albert Einstein. Criticism is ahead in chapter 7.

7. Lorentz Fitzgerald length contraction recalculated with correction of mistakes in original calculations.

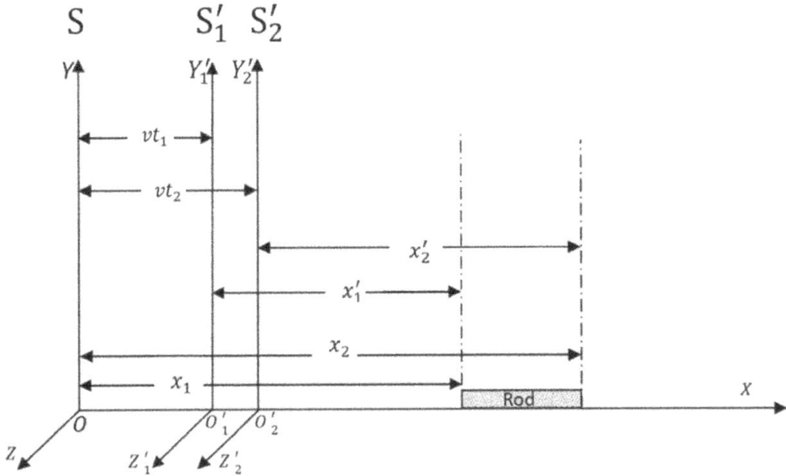

Figure: 7.1

Here a cartesian plane is aligned to the steady inertial reference frame S. S' inertial reference frame is moving with uniform rectilinear velocity "v" in the direction of an X axis. Two situations of the S' are shown to get x values for proximal and distal and of the Rod.

Rod is placed on X axis. In a stationary reference frame S, Co-ordinates of the proximal end of the Rod is $(x1, 0, 0, t1)$ and co-ordinates of the distal end is $(x2, 0, 0, t2)$

Now,

Lorentz transformation equations

$$x' = \frac{x - vt}{\sqrt{1 - \dfrac{v^2}{c^2}}} \dots\dots\dots\dots\dots\dots\dots\dots\dots\dots (9)$$

$$x = \frac{x' + vt'}{\sqrt{1 - \frac{v^2}{c^2}}} \dots\dots\dots\dots\dots\dots\dots\dots\dots\dots (10)$$

In a figure 7.1 the Rod is put along the positive X axis. Length of the rod in relation to stationary reference frame is proper length l_0 Length of the rod in relation to moving reference frame is contracted length l

Proper length $\qquad\qquad l_0 = x_2 - x_1 \qquad\qquad\qquad \dots\dots\dots (15)$

Contracted length $\qquad\quad l = x_2' - x_1' \qquad\qquad\qquad \dots\dots\dots (16)$

Now let's put value of (10) in the equation no (16).

$$l = x_2' - x_1'$$

$$l = \frac{x_2 - vt_2}{\sqrt{1 - \frac{v^2}{c^2}}} - \frac{x_1 - vt_1}{\sqrt{1 - \frac{v^2}{c^2}}}$$

$$= \frac{x_2 - vt_2 - x_1 + vt_1}{\sqrt{1 - \frac{v^2}{c^2}}}$$

$$l = \frac{(x_2 - x_1) - (vt_2 - vt_1)}{\sqrt{1 - \frac{v^2}{c^2}}}$$

$$l = \frac{(x_2 - x_1) - \left(\frac{v}{c}ct_2 - \frac{v}{c}ct_1\right)}{\sqrt{1 - \frac{v^2}{c^2}}}$$

$$l = \frac{(x_2 - x_1) - \frac{v}{c}(ct_2 - ct_1)}{\sqrt{1 - \frac{v^2}{c^2}}}$$

$$= \frac{(x_2 - x_1) - \frac{v}{c}(x_2 - x_1)}{\sqrt{1 - \frac{v^2}{c^2}}}$$

$$= \frac{l_0 - \frac{v}{c}l_0}{\sqrt{1 - \frac{v^2}{c^2}}}$$

$$= l_0 \frac{(1 - \frac{v}{c})}{\sqrt{1 - \frac{v^2}{c^2}}}$$

$$= l_0 \frac{\sqrt{1 - \frac{v}{c}}}{\sqrt{1 + \frac{v}{c}}} * \frac{\sqrt{1 - \frac{v}{c}}}{\sqrt{1 - \frac{v}{c}}}$$

$$l = l_0 \frac{\sqrt{1 - \frac{v}{c}}}{\sqrt{1 + \frac{v}{c}}} \qquad where\ v < c.$$

$$l = l_0 \sqrt{\frac{1 - \frac{v}{c}}{1 + \frac{v}{c}}} = l_0 \sqrt{\frac{\frac{c - v}{c}}{\frac{c + v}{c}}}$$

$$l = l_0 \sqrt{\frac{c - v}{c + v}}$$

Above equation shows the value of contracted length l.

Length contraction is Δl

$$\Delta l = l_0 - l$$

$$\Delta l = l_0 - l_0 \sqrt{\frac{c - v}{c + v}}$$

$$\Delta l = l_0 \left(1 - \sqrt{\frac{c-v}{c+v}}\right) \qquad \text{(where, v<c)}$$

In the Special theory of Relativity,

vt_2 and vt_1 are considered same with the argument that x_2' and x_1' are measured simultaneously, so t_2 and t_1 remains same and vt_2 becomes equal to vt_1. So, that gets cancelled during calculations.

But The phrase "x_2' and x_1' are measured simultaneously" is an impossible task for $x_2 \neq x_1$.

Let's understand this.

See the figure 5.1 in the Lorentz transformation.

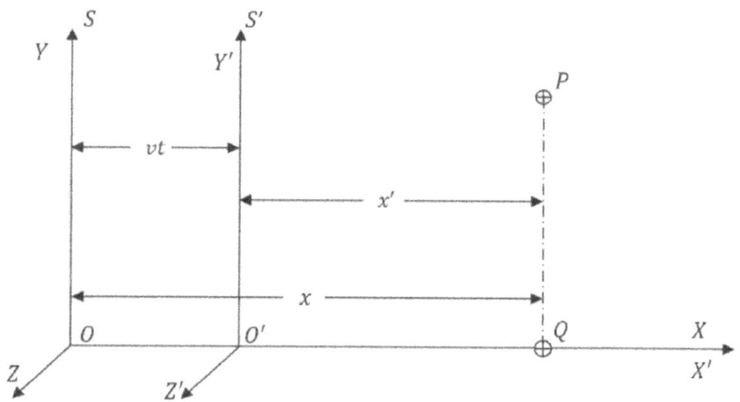

Figure: 5.1

Here, vt is the distance of Y' axis of *the S'* frame from the Y axis of the S frame.
This is the distance that reference frame S' travels along the X axis in a time t.
Now what is this time t?

This is the time taken by photon traveling with the speed c from Q to O and distance equal to the value of co-ordinate x of the point P or Q in reference frame S.

And,

$$ct = x$$

Here c is the constant, so, value of t is depending on the value of x.

As the value of x changes, value of the t will also change.

If we consider the value of v constant,

when we measure x', then vt depends on the value of x.

$$vt = xv/c$$

Means a value of x determines a value of t in a ct and that value of t is there in a vt which affects the value of vt and thus value of x' also.

So, for $x_1 \neq x_2 \Leftrightarrow vt_1 \neq vt_2 \Leftrightarrow x'_1 \neq x'_2$

Lorentz transformation equations are equations which are made to calculate co-ordinates of one point only at one time, where time $t = x/c$. If x varies, then time t will also vary according to the value of x, you can't choose that time t according to your free will for particular x and can't choose same t for two different x values.

So, we can't measure both x_2 and x_1 simultaneously. Because when we apply Lorentz transformation Equation for point x_2, the value of x_2 will be equal to ct_2 where t_2 will be specific for x_2. In that time t_2, the moving reference frame S' will move a distance vt_2, which is also a specific distance depending on the value of t_2 and so, also depending on the value of x_2.

when we apply Lorentz transformation Equation for point x_1, the value of x_1 will be equal to ct_1 where t_1 will be specific for x_1. In that time t_1 the moving reference frame S' will move a distance vt_1. which is also a specific distance depending on the value of t_1 and so, also depending on the value of x_1.

So, the value of vt_2 can't be equal to vt_1, means a position of reference frame S' can't be at same place when we are applying the Lorentz transformation equations for two different points having value of x co-ordinates different from each other.

So, measuring distance for proximal and distal end of the rod simultaneously in a moving reference frame is not possible, because when we apply Lorentz transformation equation, value of t' will be different for different x values when velocity v of the moving reference frame will remain constant. So, we can calculate value of x_1' and x_2' for respective value x_1 and x_2 at same velocity v, but we can't calculate them at the same time t.

Let's think about the length contraction described in the book "Relativity, the Special and the General theory by Albert einstein, 1916"
Following is described in that book
-:-

"There, rod is put in a moving reference frame with proximal end on the (0,0). There, time t_1 becomes 0, and time t_2 becomes t and,
contracted length is $\quad l = x_2 - 0 = x_2$

proper length is $l_0 = x_2' - x_1' = Y(x_2 - vt - x_1 + vt)$

And so,

$$l_0 = Y(x_2 - x_1)$$

$$l_0 = Y(x_2 - 0)$$

$$l_0 = Yx_2$$

$$l_0 = Yl$$

$$l = \frac{l_0}{Y}$$

$$l = l_0\sqrt{1 - \frac{v^2}{c^2}}$$

-:-

So, there we can find contracted length as above. We could not see the mistake in the calculation there because there t_1 is taken as 0. And vt is canceled anyhow by taking $t_2 = t = 0$.

The line is there, "Lorentz transformation value of both end of the rod at time $t = 0$ is as below"

Means in that case, in the book of the theory of relativity it is mentioned that the distance of the distal end of the rod placed in a moving reference frame can be calculated by Lorentz transformation equation in relation to stationary reference frame S by taking $t = 0$ value in the vt

This is the mistake, if proper distance x is not zero, then ct cannot be zero.

If ct cannot be zero, then vt cannot be zero for v more than 0. So, by using Lorentz transformation equation one cannot take the value of $t = 0$ for the point having value of x coordinate more than zero. So, mistake is also there in the Book of Relativity by albert Einstein.

Please note that I am not agree with the corrected equations shown above for Lorentz Fietz Gerald's length contraction because as described ahead there is a wrong calculation in the derivation of the constant Y.

What could be the proper equation for relativistic length contraction by accepting both postulates of the theory of the special relativity?

It could be as follow.

relativistic transformation equations are,

$$x' = t(c - v) \quad \dots\dots\dots\dots (13)$$
$$y' = y$$
$$z' = z$$
$$t' = t\frac{c - v}{c} \quad \dots\dots\dots\dots. (14)$$

So, here,

Proper length $\qquad l_0 = x_2 - x_1$ $\qquad\qquad$ ... ... (15)

Contracted length $\qquad l = x_2{}' - x_1{}'$ $\qquad\qquad$ ... ... (16)

Now let's put value of (13) in the equation no (16)

$$l = x_2{}' - x_1{}'$$
$$l = t_2(c - v) - t_1(c - v)$$
$$= t_2 c - t_2 v - t_1 c + t_1 v$$
$$l = ct_2 - ct_1 - (vt_2 - vt_1)$$
$$l = (x_2 - x_1) - (vt_2 - vt_1)$$
$$l = l_0 - (vt_2 - vt_1)$$

Above equation shows the value of contracted length l.

Length contraction is Δl

$$\Delta l = l_0 - l$$
$$\Delta l = l_0 - (l_0 - (vt_2 - vt_1))$$
$$\Delta l = v(t_2 - t_1) \qquad\qquad (here, v < c)$$

This could be the proper calculation for the relativistic apparent length contraction in a case where moving reference frame is moving toward the source of the light or observed subject.

There would be a length elongation in case of the moving observer is moving away from the observed subject or the source of the light.

So, as I have cleared points above the time dilation equation described in the Special theory of Relativity is wrong and there should be a time contraction when observer moves away from the observed thing. Similarly, there should be length elongation also if we consider observer specific constancy of the light true. If we consider Aether as an existence, then the Relativity in relation to the Aether would be different From the Special theory of Relativity.

8.Photon clocks

Let me cover the photon clock topic.

I have told ahead in this chapter that using photon clock example to show a time dilation is inappropriate. This is explained below.

There are two types of photon clocks described below. Photon clock type 1 is a classical photon clock used in the Special theory of Relativity where it shows example of the relativistic time dilation with simple calculations while photon clock type 2 is an assembly that enables us to derive a time contraction which is not described in the special theory of the relativity.

Photon clock type 1

Photon clock type 1 explains the time dilation concept in a simple manner.

(This chapter is here to understand an example of photon clock given by albert Einstein in support of his special theory of relativity. This chapter is essential to understand photon clock type II explained ahead in book. This chapter is given from the perspective of special relativity theory. Author don't agree with all these.)

Here showing a figure of a photon clock type1.

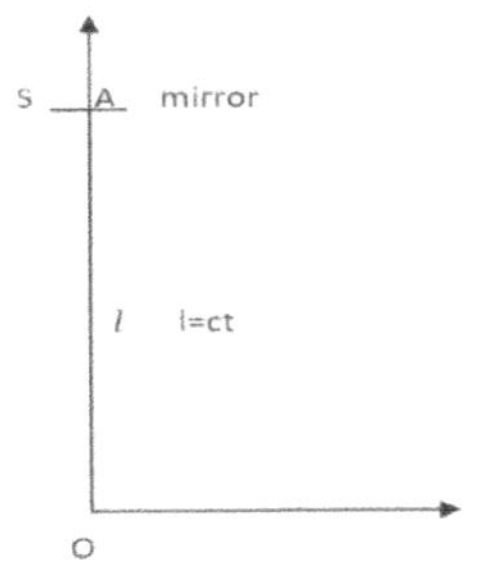

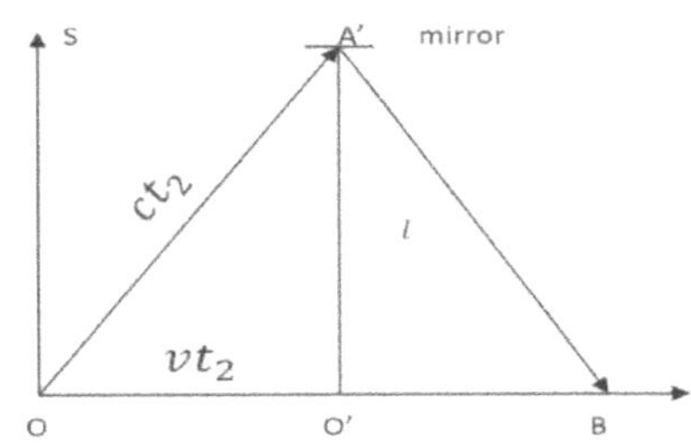

Situation A
When Clock is not moving.
Light path is small.
Time in stationary clock is faster.

Situation B
When clock is moving with velocity v.
Light path increases in length.
Time dilates in moving reference frame.

Figure: 8.1

(Simply from these figures, we can say that light path for the stationary observer increases when clock is moving in relation to stationary observer, speed of light c has to remains same. So, for compensation, time has to get slow down or has to be dilated)

Please understand here that, t_2 is the dilated or slowed down time, means it is the time what clock takes to complete one half tick when moving. Means if time dilates then t_2 will be greater than t. While stationary clock will complete 1 tick for 1 second, there will be less than one second in moving clock.

Here,

As shown in Situation A, S is a stationary inertial reference frame. Clock is stationary with reference to S and also stationary in relation to observer. Light travels from O to A, strikes mirror at A and come back to O. Time to travel on one directional path for light is t.
Photon travels distance $l = OA$ in a time t.
Speed of light is c,
So,

$$l = ct \quad \dots\dots\dots\dots\dots\dots\dots\dots\dots\dots\dots. (p.1)$$

Now,

As shown in Situation B.

S is stationary reference frame and photon clock is moving with velocity v along the positive X axis in reference to stationary observer. Now stationary observer from S reference frame will see the situation as shown in figure for situation B. Light travels from O to A', strikes mirror at A' and come back to B. Time in moving clock is t_2 in which photon travels OA' distance in situation B. Photon travels distance $OA' = ct_2$ in a time t_2. Clock moves a distance OO' in a time t_2 with the uniform rectilinear velocity v along with X axis. t_2 is a time that photon takes to move half of the path. Value of t_2 will always be greater than value of t. So, it can be said that the clock in moving reference frame will show less time compared to clock at rest.
So,

$$OO' = vt_2$$

$$OA' = ct_2$$

Applying Pythagoras theorem.

$$OA'^2 = OO'^2 + O'A'^2$$

$$c^2 t_2{}^2 = v^2 t_2{}^2 + l^2$$

Putting value of l from equation no (p.1)

$$c^2 t_2{}^2 = v^2 t_2{}^2 + l^2$$

$$c^2 t_2{}^2 = v^2 t_2{}^2 + c^2 t^2$$

$$t_2{}^2 (c^2 - v^2) = c^2 t^2$$

$$t_2{}^2 = \frac{c^2 t^2}{(c^2 - v^2)}$$

$$t_2{}^2 = \frac{t^2}{\frac{(c^2 - v^2)}{c^2}}$$

$$t_2{}^2 = \frac{t^2}{\left(1 - \frac{v^2}{c^2}\right)}$$

$$t_2 = \frac{t}{\sqrt{1 - \frac{v^2}{c^2}}}$$

As we know that $1 \Big/ \sqrt{1 - \frac{v^2}{c^2}}$ is a factor Y,

$$t_2 = Yt.$$

Let me show an example to understand the dilation of the time.

If clock is moving with the velocity equal to $c/2$ means with the speed half of the speed of the light,

Then,

$$v = \frac{c}{2}$$

$$t_2 = \frac{t}{\sqrt{1 - \frac{(c/2)^2}{c^2}}}$$

$$t_2 = \frac{t}{\sqrt{1 - \frac{c^2}{4c^2}}}$$

$$t_2 = \frac{t}{\sqrt{1 - \frac{1}{4}}} = \frac{t}{\sqrt{0.75}} = 1.16t$$

$$t_2 = 1.16t$$

Moving clock will take more time to tick and it will be slower.

$$t = \frac{t_2}{1.16} = 0.86$$

This means for every second on the stationary clock moving clock will pass 0.86 seconds.

Photon clock Type-2

Now, I would like to show a photon clock type 2 that can show time contraction.

Photon clock Type-2 illustration.

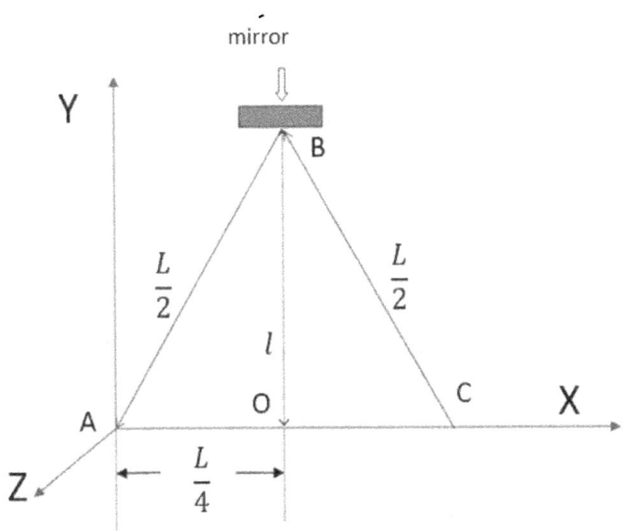

Figure 8.2

Here is the illustration of the photon clock type 2. Light source is at C, from where photon arises and travels toward B, where it gets reflected by mirror, and moves to reach to the point A. Total Distance travelled by light in a time t in nonmoving clock for stationary observer is CBA= $L = ct$. Angle BCA is 60°, angle BAC and angle ABC are also of the 60°. Angle BOA is right angle. So, Distances,

AB=BC=AC=ct/2=L/2.

Distance AO=1/2 *AC=L/4.

Height of the clock is l which is the straight minimum distance between X axis and mirror. Motion of light is two dimensional and on the path CBA.

Here, on the next page two reference frames are illustrated in a figure 8.3.

Reference frame S is a situation A

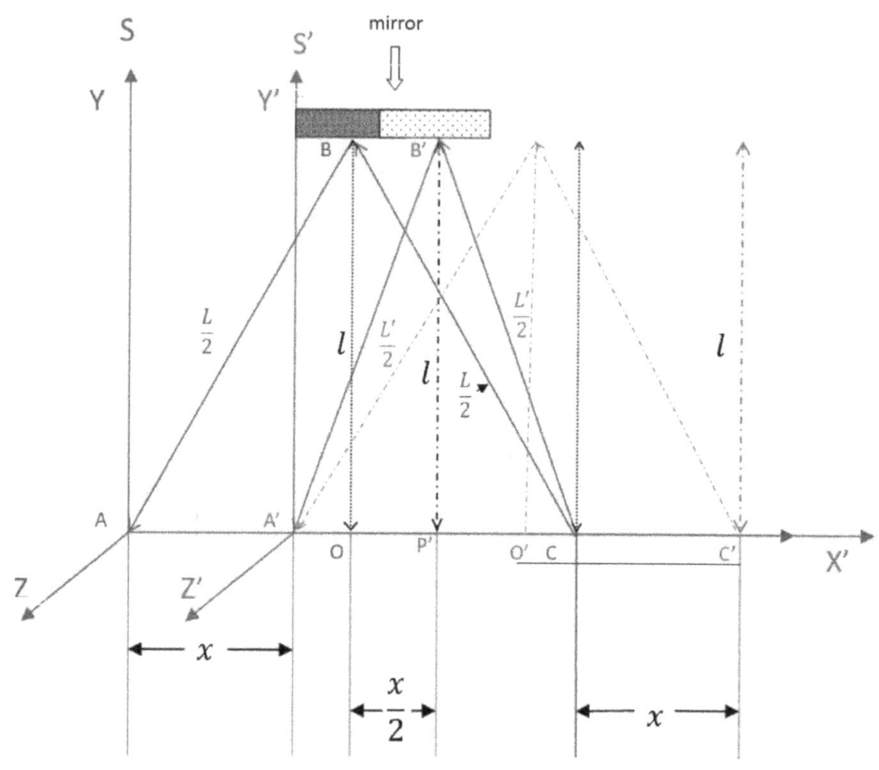

Figure 8.3

In a situation A, photon clock is not moving in reference to stationary observer in reference frame S. CBA is the path of photon, which photon completes in a time t. Length CBA is equal to L. $L = ct$. O is the point on X axis that show the distance of B from Y axis. B is the point at where mirror is placed, and photon get reflected from this point at the height l towards the point A.

OB= l .

So, here height of the photon clock is l which will remain same even when the clock will move in the case of situation B. The path of the photon makes a perfect triangle along with X axis, with each corner made up of 60°, So in a figure 8.3

AB=BC=AC=2AO=2OC=$L/2$

So, AO=$L/4$

Light travels from C to B and then B to A. Total length of light path from C to A, via B is L, which light travels in a time t with the constant speed of light c.

So, Distance $CBA = L = ct.$

Distance $AO = L/4 = ct/4$

Let's find out the relation between l and ct.

In right triangle AOB, $\llcorner$AOB is the right angle.

So, according to Pythagoras theorem.

$$AB^2 = AO^2 + OB^2$$

$$\left(\frac{L}{2}\right)^2 = \left(\frac{L}{4}\right)^2 + l^2$$

$$\frac{L^2}{4} = \frac{L^2}{16} + l^2$$

$$\frac{L^2}{4} - \frac{L^2}{16} = l^2$$

$$l^2 = \frac{4L^2 - L^2}{16}$$

$$l^2 = \frac{3L^2}{16}$$

$$l = \frac{\sqrt{3}}{4}L$$

$$l = \frac{\sqrt{3}}{4}ct$$

$$ct = \frac{4l}{\sqrt{3}}$$

Now, Let's think about situations B.

Situations B.

      In situation B, Photon clock with its reference frame S' moves with velocity v along with the Positive X axis. It travels a distance x in a time t with the velocity v along the Positive X axis.

So,

$x = vt = AA' = CC'$.

      This is the distance that clock moves in time t on positive X axis with the velocity v. Photon travels the distance L' in a time t_2 which is equal to ct_2 when speed of the photon is c. Here time is denoted as t_2 and not t, because photon has an observer specific constant speed, and here when clock moves, the path distance of photon for the stationary observer becomes short and to keep speed of the light same there would be a time contraction here which is shown as a time t_2. Here photon travels less distance for the stationary observer in a situation B compared to a distance traveled by photon in a situation A. You can see this path difference in the figure 8.3. You can see in a figure that base of the triangle formed by photon path in a moving clock appears contracted for the stationary observer in S frame. Length $A'C$ is less than AC. The difference is equal to vt which is equal to $AA' = CC'$.

So, length of the base of the triangle in a situation B will be $A'C$,

So,

$$A'C = AC - vt$$
$$A'C = \frac{ct}{2} - vt$$

      So, in a situation B, height of the photon clock doesn't change, and it remains l.

Here, right triangle $A'B'P'$ formed with right angle $\llcorner A'P'B'$.

$A'P'$ is half of the $A'C$.

$$A'P' = \frac{1}{2}A'C = \frac{1}{2}\left(\frac{ct}{2} - vt\right)$$

$$A'P' = \frac{ct}{4} - \frac{vt}{2}$$

according to Pythagoras theorem.

$$(A'B')^2 = (A'P')^2 + (B'P')^2$$

$$\left(\frac{L'}{2}\right)^2 = \left(\frac{ct}{4} - \frac{vt}{2}\right)^2 + l^2$$

$$\left(\frac{ct_2}{2}\right)^2 = \left(\frac{ct}{4} - \frac{vt}{2}\right)^2 + l^2$$

Let's put the value of l^2 .

$$l^2 = \frac{3c^2t^2}{16}$$

So,

$$\left(\frac{ct_2}{2}\right)^2 = \left(\frac{ct}{4} - \frac{vt}{2}\right)^2 + \frac{3c^2t^2}{16}$$

$$\frac{c^2t_2{}^2}{4} = \left(\frac{c^2t^2}{16} - \frac{ctvt}{4} + \frac{v^2t^2}{4}\right) + \frac{3c^2t^2}{16}$$

$$\frac{c^2t_2{}^2}{4} = \frac{1}{4}\left[\left(\frac{c^2t^2}{4} - \frac{ctvt}{1} + \frac{v^2t^2}{1}\right) + \frac{3c^2t^2}{4}\right]$$

$$c^2t_2{}^2 = \left(\frac{c^2t^2}{4} - \frac{ctvt}{1} + \frac{v^2t^2}{1}\right) + \frac{3c^2t^2}{4}$$

$$c^2t_2{}^2 = \frac{c^2t^2}{4} - ctvt + v^2t^2 + \frac{3c^2t^2}{4}$$

$$c^2t_2{}^2 = c^2t^2 - ctvt + v^2t^2$$

$$t_2{}^2 = t^2 - \frac{vt^2}{c} + \frac{v^2t^2}{c^2}$$

$$t_2{}^2 = t^2 \left(1 - \frac{v}{c} + \frac{v^2}{c^2} \right)$$

$$t_2 = t \sqrt{\frac{v^2}{c^2} - \frac{v}{c} + 1} \ \dots\dots\dots\dots\dots\dots\dots\dots\dots\dots\dots \ (Q)$$

(We neglected negative value for time here because time don't move backward here.)

This is the relationship between variated time t_2 with time t in stationary clock.

Here path contracts, So, time will also contract for compensation to keep velocity c constant when clock is moving with velocity v.

So, for each second in a stationary clock, time passing in a moving clock would be more than a second, means moving clock will tick faster.

So, here t_2 will be smaller than t in contrast to Type-1 Photon clock, where time dilates as shown in a topic just before this topic.

t_2 is a time taken to complete 1/2 tick. When it is smaller, clock shows more time and when it is bigger clock shows less time.

This was a calculations according to the Special theory of Relativity and this prediction is not mentioned anywhere in the special theory of the relativity.

Here the point is, if there is a moving inertial reference frame and a light moving with its source in that reference frame and there is a stationary observer in a stationary inertial reference frame then it is not always necessary that there will be an elongated path of that light for the stationary observer compared to path observed by observer in moving reference frame. reverse is also possible as shown in this Photon clock type 2 example. Assembly of the photon clock is a theoretical and we just need a half tick so reflecting back of the light photon for further tick is not required here, but it can be achieved by putting vertical mirror glass at a point A and putting another mirror below at a distance l which should be aligned with the mirror at B to reflect light back again at the source. So, according to the relative direction of the light beam there can be a variation on the length of the observed path by stationary observer when source of the light is moving in relation to that observer. Application of the single time dilation equation is not possible here for all possibilities depending on the direction of the light beam. So, example of the photon clock to show that time dilation is wrong.

[End of the book]

www.ingramcontent.com/pod-product-compliance
Lightning Source LLC
Chambersburg PA
CBHW040111150726
48005CB00013B/1662